PERENNIAL MATH

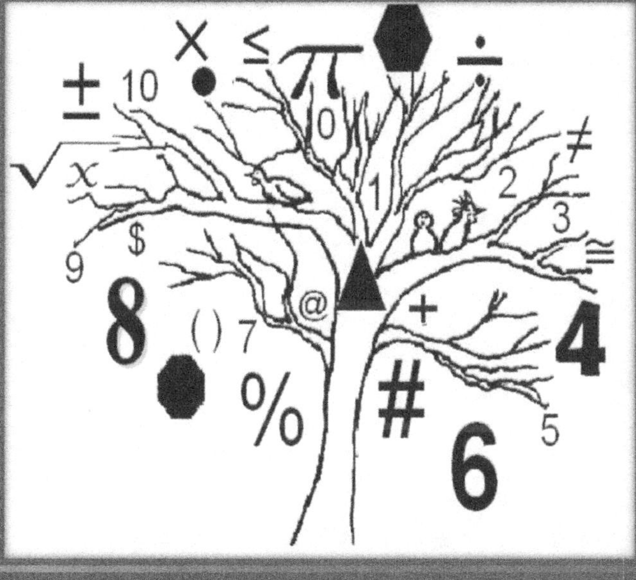

CONTEST PROBLEMS

2011-2014

Dr. Sylvia Dean has been a gifted specialist for 25 years in Huntsville, Alabama. Dr. Dean created Perennial Math in 2011 upon completion of her doctoral degree. She has hosted many robotics, technology, and mathematics competitions throughout the years and has tirelessly worked on providing educational opportunities to gifted students around the world. It has always been her ambition to leave students with the feeling that "math is cool."

These problems are challenging and students should know up front that scoring a 3 out of 6 on any of these tests is extraordinary and above average. Great problem-solvers learn persistence and appreciation for challenge. Good luck as you encourage your students to take the quest to become great mathletes.

INTERMEDIATE LEVEL
Grades 5-6

PERENNIAL MATH

TEST # 1

November 2011 Intermediate

1) Counting from either end of the line, Suzy is the 13th person in line. How many people are in the line?	
2) The average of 5 numbers is 6. Two of the numbers are 8 and 7. The other three numbers are equal. What is the value of one of the three equal numbers?	
3) A baby bird chirps 4 times every 5 seconds. At this rate, how many times will the baby bird chirp in 5 minutes?	
4) There are 2 girls for every boy in the 6th grade. Out of the 72 total students in the 6th grade, 7 of the boys have braces. How many of the boys do not have braces?	
5) Rectangle *ABCD* is split into 4 smaller rectangles. Each side of each rectangle is a whole number of centimeters. The areas of three of the smaller rectangles are shown. What is the area of rectangle *ABCD*? 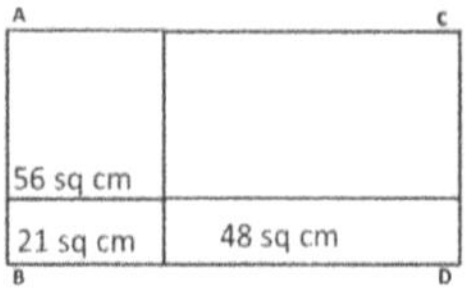	

PERENNIAL MATH

TEST # 1

November 2011 Intermediate Answers

<table>
<tr><td>1) Counting from either end of the line, Suzy is the 13th person in line. How many people are in the line?</td><td>25</td></tr>
<tr><td>2) The average of 5 numbers is 6. Two of the numbers are 8 and 7. The other three numbers are equal. What is the value of one of the three equal numbers?</td><td>5</td></tr>
<tr><td>3) A baby bird chirps 4 times every 5 seconds. At this rate, how many times will the baby bird chirp in 5 minutes?</td><td>240</td></tr>
<tr><td>4) There are 2 girls for every boy in the 6th grade. Out of the 72 total students in the 6th grade, 7 of the boys have braces. How many of the boys do not have braces?</td><td>17</td></tr>
<tr><td>5) Rectangle ABCD is split into 4 smaller rectangles. Each side of each rectangle is a whole number of centimeters.
The areas of three of the smaller rectangles are shown. What is the area of rectangle ABCD?
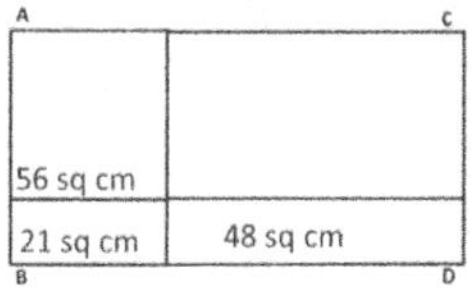
</td><td>$253cm^2$</td></tr>
</table>

PERENNIAL MATH TEST # 1

November 2011 Intermediate Solutions

1) There are 12 people ahead of Suzy and 12 people behind her. 12 + 12 + Suzy = **25**.

25

2) If the average of 5 numbers is 6, then the sum of those numbers is 5 x 6, or 30. If you subtract the sum of the two known numbers 8 and 7 or 15 from 30, you have 15. The three equal numbers add up to 15. The value of one of the equal numbers is 15/3 or **5**.

5

3) Each minute has 60 seconds. So if the bird chirps every 5 seconds, there are 12 chirping intervals each minute. If the bird chirps 4 times each interval then there are 12 x 4, or 48 chirps per minute. The bird chirps 48 x 5 or **240 chirps** in 5 minutes.

240

4) The ratio of girls to boys is 2:1. If there are 72 total students, then there is a 48:24 ratio of girls to boys. If 7 boys wear braces, then 24-7 or **17 boys** do not wear braces.

24 girls	24 girls	24 boys

24 total boys
<u>-7 boys with braces</u>
17 boys without braces

17

PERENNIAL MATH

TEST # 1

November 2011 Intermediate Solutions

5) Rectangle *ABCD* is split into 4 smaller rectangles. The small rectangle with area 21 cm^2 must have sides that are 3 cm by 7 cm. Since this rectangle shares a side with the other two rectangles whose areas are 56 cm^2 and 48 cm^2, its dimensions will be used to solve for the other rectangles' dimensions. The dimensions of the rectangle with 56 cm^2 must be 7 cm by 8 cm. The dimensions of the rectangle with 48 cm^2 must be 3 cm by 16 cm. The dimensions of the unknown rectangle must be 16 cm x 8 cm; so the area of the unknown rectangle is 128 cm^2. The area of rectangle *ABCD* is equal to the sum of all the smaller rectangles 128cm^2 + 56 cm^2 + 48 cm^2 + 21 cm^2=253 cm^2; or A=length x width, 23 cm x 11cm = **253 cm^2**.

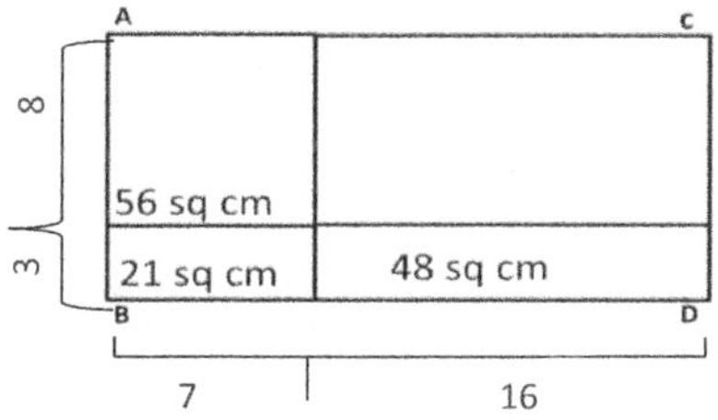

253 cm^2

PERENNIAL MATH

TEST # 2

December 2011 Intermediate

1) What number should be written in the box to make the statement true? 16 x 48 = 32 x ☐	
2) The four-digit number 8C5C is divisible by 9. What digit does the C represent?	
3) Pedro earns $4 for raking leaves each day. He can earn $7 instead for doing additional tasks. At the end of 12 days Pedro has earned a total of $60. On how many of these days did Pedro do additional tasks?	
4) This tower is made of layers of congruent cubes. If two more layers were built below the bottom layer following the increasing pattern, how many cubes would make up this figure? 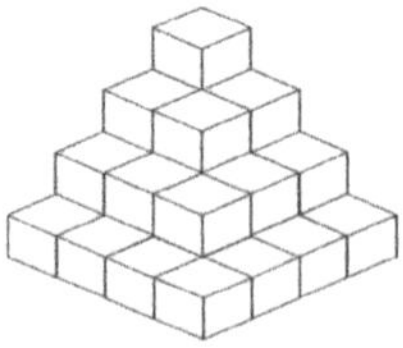	
5) Amy's mother is 4 times as old as Amy is now. In 4 years, Amy's mother will be 3 times as old as Amy will be then. How old was Amy's mother when Amy was born?	

PERENNIAL MATH

TEST # 2

December 2011 Intermediate Answers

1) What number should be written in the box to make the statement true? 16 x 48 = 32 x □	**24**
2) The four-digit number 8C5C is divisible by 9. What digit does the C represent?	**7**
3) Pedro earns $4 for raking leaves each day. He can earn $7 instead for doing additional tasks. At the end of 12 days Pedro has earned a total of $60. On how many of these days did Pedro do additional tasks?	**4**
4) This tower is made of layers of congruent cubes. If two more layers were built below the bottom layer following the increasing pattern, how many cubes would make up this figure? 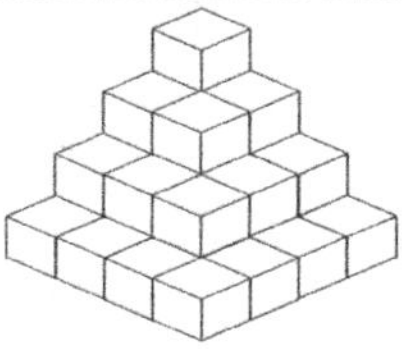	**91**
5) Amy's mother is 4 times as old as Amy is now. In 4 years, Amy's mother will be 3 times as old as Amy will be then. How old was Amy's mother when Amy was born?	**24**

PERENNIAL MATH

TEST # 2

December 2011 Intermediate Solutions

1) 16 x 48 = 768. Since both sides of the equation are equal, divide 768 by 32 and the number that should go in the box is **24**.

24

2) In order for a number to be divisible by 9, the sum of its digits must be divisible by 9. Since the sum of the digits of 8C5C is 13, then the next multiple of 9 is 18, but it is not possible to get the sum of 18 adding 2C. So the next multiple of 9 is 27. 2C can add up to 14. C must be **7**.

7

3) Make a chart to see how much money Pedro could earn in 12 days. Pedro did extra tasks for **4 days**.

Days	1	2	3	4
$7	7	14	21	28
$4	44	40	36	32
Days	11	10	9	8
Total	51	54	57	60

4

4) Some blocks are hidden, but they must be there in order to support the blocks on top of them. The original tower has layer of 1, 4, 9, and 16 blocks. If the pattern were to be continued in an increasing manner, then the next layers would also be square numbers. The next two layers would be 25 and 36 blocks. 1 + 4 + 9 + 16 + 25 + 36 = **91 blocks**.

91

PERENNIAL MATH

TEST # 2

December 2011 Intermediate Solutions

5) Make a chart to see what Amy's age is compared to mom's and make rows for their ages in 4 years from now. Look for numbers that will be 3 times the amount. The answer is that Amy is 8 years old now and her mother is 4 times that or 32 years old. In 4 years she will be 12 and her mother will be three times as old or 36 years old. Subtract 36-12 and Amy's mother was **24 years old** when Amy was born.

Mom's age now	4	8	12	16	20	24	28	32
Amy age now	1	2	3	4	5	6	7	8
Mom's age in 4 years	8	12	16	20	24	28	32	36
Amy's age in 4 years	5	6	7	8	9	10	11	12

24

PERENNIAL MATH TEST # 3

January 2012 Intermediate

1) What are the greatest number of Wednesdays that can occur in 48 days?	
2) Sandy went shopping and paid \$30 for a necklace and received \$7 change. She paid another \$15 for a bracelet and received \$4 change. She had \$22 in her wallet after shopping. How many dollars did she have in her wallet after she bought the necklace?	
3) Circles A, B, and C form four regions where they intersect. The numbers 1, 3, 4, and 6 can be placed in those regions so that all circles have the same sum. What is that sum? 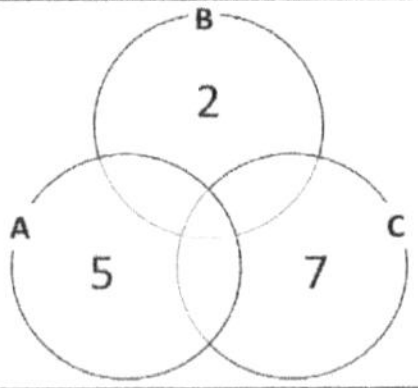	
4) The area of a rectangle is 60 square centimeters. The lengths of the sides are whole numbers. What is the smallest perimeter in centimeters this rectangle can have?	
5) A kiwi and a peach costs 27 cents, an orange and kiwi costs 19 cents, and a peach and orange costs 24 cents. How many cents does it cost for one kiwi, one peach, and one orange?	

PERENNIAL MATH

TEST # 3

January 2012 Intermediate Answers

1) What are the greatest number of Wednesdays that can occur in 48 days?	**7**
2) Sandy went shopping and paid $30 for a necklace and received $7 change. She paid another $15 for a bracelet and received $4 change. She had $22 in her wallet after shopping. How many dollars did she have in her wallet after she bought the necklace?	**33**
3) Circles A, B, and C form four regions where they intersect. The numbers 1, 3, 4, and 6 can be placed in those regions so that all circles have the same sum. What is that sum? 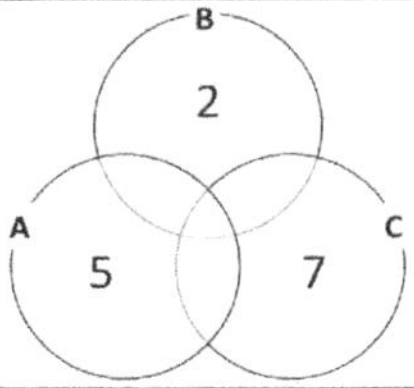	**15**
4) The area of a rectangle is 60 square centimeters. The lengths of the sides are whole numbers. What is the smallest perimeter in centimeters this rectangle can have?	**32**
5) A kiwi and a peach costs 27 cents, an orange and kiwi costs 19 cents, and a peach and orange costs 24 cents. How many cents does it cost for one kiwi, one peach, and one orange?	**35**

PERENNIAL MATH

TEST # 3

January 2012 Intermediate Solutions

1) The greatest number of Wednesdays that can occur in 48 days is **7** because if you started a calendar on a Wednesday, then day 1, 8, 15, 22, 29, 36, and 43 would all be Wednesdays .

7

2) Sandy spent (30-7) or \$23 on the necklace. Then she also spent (15-4) or \$11 on the bracelet. She had \$22 in her wallet when done shopping. \$22 left + \$34 spent = \$56 when she started shopping. If she had \$56 when she started shopping and spent \$23 at the first store she would have had **\$33** in her wallet after she bought the necklace.

33

3) The numbers 1, 3, 4, and 6 can be placed in the four regions where Circles A, B, and C intersect so that all circles have the same sum of **15**.

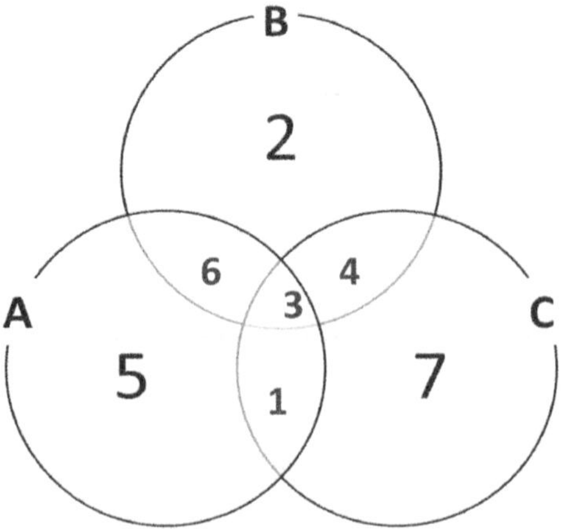

15

PERENNIAL MATH

TEST # 3

January 2012 Intermediate Solutions

4) First, find all the rectangles with an area of 60 square centimeters whose lengths of their sides are whole numbers. Next, add up the dimensions to find the perimeter P=2 (length + width). The smallest perimeter is 2 (10 + 6) = **32cm**.

Length	Width	Perimeter
60	1	122
30	2	64
20	3	46
15	4	38
12	5	34
10	6	32

32

5) Use algebra to determine the cost of one kiwi + one peach + one orange. A kiwi and a peach costs 27 cents, an orange and kiwi costs 19 cents, and a peach and orange costs 24 cents. If you add all the fruit together you get 2 of each kind which adds up to 70 cents. So half of that would be 1 of each fruit and it would cost half as much or **35 cents**.

		K	+	P	=	27
O	+	K			=	19
O	+			P	=	24
2O	+	2K	+	2P		70

2O + 2K + 2P/2= 70 cents/2

O + K + P = **35 cents**.

35

PERENNIAL MATH TEST # 4

February 2012 Intermediate

1) Hank and Calvin each have the same amount of tickets. Hank buys 1 lollipop with his tickets and still has 90 tickets left. Calvin buys 4 lollipops and has 30 tickets left. How many tickets did Hank start with?	
2) Two years ago my age was a multiple of 7. Last year it was a multiple of 5. I am younger than 50. How old am I now?	
3) What is the total number of rectangles of all sizes that can be found in this diagram? 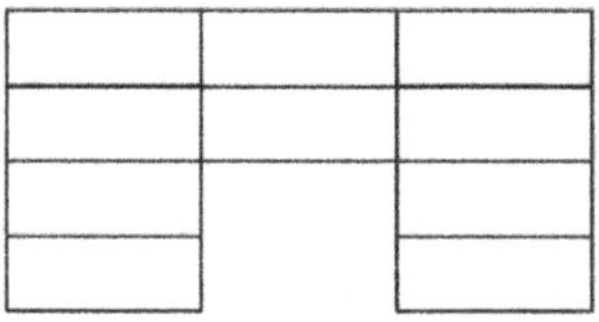	
4) Mike has 10 tennis balls. They have weights of 2g, 4g, or 5g. He has a different number of balls of each weight. What is the largest possible weight or total grams of Mike's tennis balls?	
5) Numbers such as 654 and 631 have digits that are in decreasing order. The number 644 does not have digits in decreasing order. How many numbers between 100 and 1000 have their digits in decreasing order?	

PERENNIAL MATH

TEST # 4

February 2012 Intermediate Answers

1) Hank and Calvin each have the same amount of tickets. Hank buys 1 lollipop with his tickets and still has 90 tickets left. Calvin buys 4 lollipops and has 30 tickets left. How many tickets did Hank start with?	**110**
2) Two years ago my age was a multiple of 7. Last year it was a multiple of 5. I am younger than 50. How old am I now?	**16**
3) What is the total number of rectangles of all sizes that can be found in this diagram? 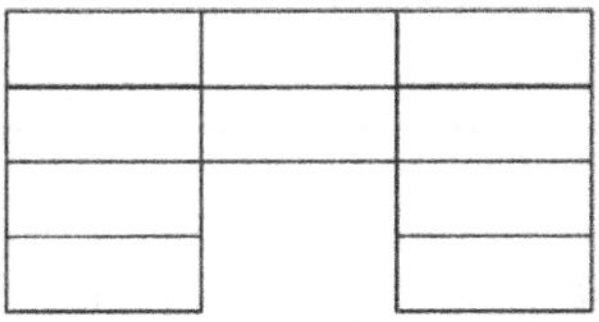	**32**
4) Mike has 10 tennis balls. They have weights of 2g, 4g, or 5g. He has a different number of balls of each weight. What is the largest possible weight or total grams of Mike's tennis balls?	**45**
5) Numbers such as 654 and 631 have digits that are in decreasing order. The number 644 does not have digits in decreasing order. How many numbers between 100 and 1000 have their digits in decreasing order?	**120**

PERENNIAL MATH

TEST # 4

February 2012 Intermediate Solutions

1) Calvin bought 3 more lollipops and ended with 60 tickets less. The lollipops must cost 20 tickets each since 60/3 = 20.

110 tickets – 20 tickets = 90 tickets

110 tickets – 80 tickets = 30 tickets.

Since the lollipops cost 20 tickets then Hank had **110 tickets** to start with.

110

2) One way to solve this to create a table to get a perspective of numbers that are 1 year and 2 years less than your current age

Age 2 Years Ago	7		14		21	28		35	42
Age Last Year	5	10	15	20	25	30	35	40	45
Age Now	6	11	16	21	26	31	36	41	46

16

3) Count the rectangles going across and then down and make a table of the different types: Rectangles made of 1 shape, 2 shapes, 3 shapes, 4 shapes, and 6 shapes. There are 32 total rectangles.

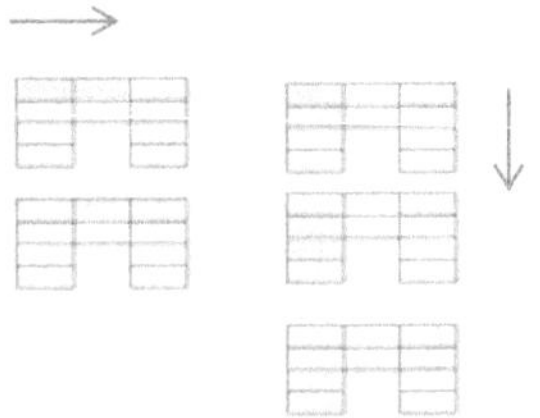

Number of rectangles in shape	How many of that type?
1	10
2	11
3	6
4	4
6	1
Total	**32**

32

PERENNIAL MATH

TEST # 4

February 2012 Intermediate Solutions

4) Start with the most of amount of the heaviest balls and add the most amount of the next highest weight, but make sure that you have a different amount of balls. Mike has 7of the 5 gram ball, 2 of the 4 gram balls, and 1 of the 2 gram balls. 7(5) + 2(4) + 2 = 35 + 8 + 2 = **45 grams.**

45

5) There is a pattern beginning with each level, the number of decreasing numbers increases by 1 more each level. There are **120** decreasing numbers.

Range	The Numbers in that Range	Total Numbers
100-200		0
201-300	210	1
301-400	310, 321, 320	3
401-500	410, 421, 420, 432, 431, 430	6
501-600	510, 521, 520, 532, 531, 530, 543, 542, 541, 540	10
601-700	610, 621, 620, 632, 631, 630, 643, 642, 641, 640, 654, 653, 652, 651, 650	15
701-800	710. 721, 720, 732, 731, 730, 743, 742, 741, 740, 754, 753, 752, 751, 750, 765, 764, 763, 762, 761, 760	21
801-900	810, 821, 820, 832, 831, 830, 843, 842, 841, 840, 854, 853, 852, 851, 850, 865, 864, 863, 862, 861, 860, 876, 875, 874, 873, 872, 871, 870	28
901-1000	910, 921, 920, 932, 931, 930, 943, 942, 941, 940, 954, 953, 952, 951, 950, 965, 964, 963, 962, 961, 960, 976, 975, 974, 973, 972, 971, 970, 987, 986, 985, 984, 983, 982, 981,980	36
	Total	**120**

↘ 1
↘ 2
↘ 3
↘ 4
↘ 5
↘ 6
↘ 7
↘ 8

120

PERENNIAL MATH

TEST # 5

March 2012 Intermediate

1) Solve: $(10 - 3)^2 + (5 + 20 \div 5) =$

2) On the basketball team, Craig who is 2 inches taller than Trevor, is 4 inches shorter than Tom. Bob is one inch taller than Dan, but three inches shorter than Craig. Trevor is 66 inches tall. How many inches tall is the tallest person on the team?

3) Clark's bike lock code is a three-digit multiple of 7 with no digits repeated. The hundreds digit is the sum of the tens digit and the ones digit. The sum of all three digits is 18. What is largest number that Clark's bike code can be?

4) There are 8 boys in a volleyball class. How many different 2-boy teams can be formed from the boys in the class?

5) The numbers 5, 6, 7, and 8 can be placed in the empty squares so that each row, column, and diagonal contains each number only once. What will be the sum of the numbers placed in squares X and Y?

X	8		5
Y			6
			7
	5		8

PERENNIAL MATH

March 2012 Intermediate Answers

TEST # 5

<table>
<tr><td>1) Solve: $(10 - 3)^2 + (5 + 20 \div 5) =$</td><td>58</td></tr>
<tr><td>2) On the basketball team, Craig who is 2 inches taller than Trevor, is 4 inches shorter than Tom. Bob is one inch taller than Dan, but three inches shorter than Craig. Trevor is 66 inches tall. How many inches tall is the tallest person on the team?</td><td>72</td></tr>
<tr><td>3) Clark's bike lock code is a three-digit multiple of 7 with no digits repeated. The hundreds digit is the sum of the tens digit and the ones digit. The sum of all three digits is 18. What is largest number that Clark's bike code can be?</td><td>945</td></tr>
<tr><td>4) There are 8 boys in a volleyball class. How many different 2-boy teams can be formed from the boys in the class?</td><td>28</td></tr>
<tr><td>5) The numbers 5, 6, 7, and 8 can be placed in the empty squares so that each row, column, and diagonal contains each number only once. What will be the sum of the numbers placed in squares X and Y?
<table>
<tr><td>X</td><td>8</td><td></td><td>5</td></tr>
<tr><td>Y</td><td></td><td></td><td>6</td></tr>
<tr><td></td><td></td><td></td><td>7</td></tr>
<tr><td></td><td>5</td><td></td><td>8</td></tr>
</table></td><td>11</td></tr>
</table>

PERENNIAL MATH

TEST # 5

March 2012 Intermediate Solutions

1) Using orders of operations PEMDAS (parenthesis, exponents, multiplication, division, addition and subtraction),

$$(10 - 3)^2 + (5 + 20 \div 5) =$$
$$(7)^2 + (5 + 4) =$$
$$49 + 9 = \mathbf{58}$$

58

2) Make a number line to represent the heights of the players. Since Craig who is 2 inches taller than Trevor, is 4 inches shorter than Tom, draw marks with Tom 4 marks ahead of Craig and Trevor 2 marks behind Craig. Since Bob is one inch taller than Dan, but three inches shorter than Craig, mark Bob three marks behind Craig (who is already marked) and place Dan 1 mark behind Bob. Put a dot on Trevor and mark it 66 inches tall. That makes Tom the tallest person on the team and he is 6 marks ahead of Trevor, so he must be **72** inches tall.

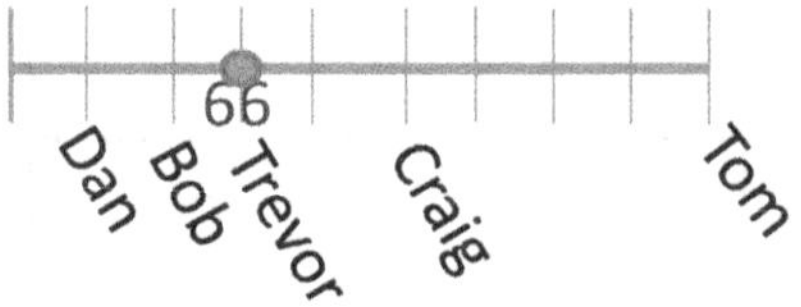

72

3) There are numerous approaches to this problem. However, we will focus on finding the largest number which is divisible by 7 and whose 3 digits adds up to 18. If 9 were the hundreds digit, the other two digits would also add up to 9 making the sum of all three digits 18. The only possible one/tens digit combinations that add up to 9 are 81, 72, 63, 54. If you try out all the possible combinations only 945 is divisible by 7. Clark's bike lock code is **945**.

Also you could use the divisibility rule of 7 which is to double the ones digit 5 and subtract it from the other remaining digits and that number would be divisible by 7 : (94-10=84 which is a number divisible by 7).

945

PERENNIAL MATH — TEST # 5

March 2012 Intermediate Solutions

4) Make a chart to show each boy and who they can pair up with and you get:
7 + 6 + 5 + 4 + 3 + 2 + 1 = **28 possible combinations**.

Boy A	Boy B	Boy C	Boy D	Boy E	Boy F	Boy G
Boy B	Boy C	Boy D	Boy E	Boy F	Boy G	Boy H
Boy C	Boy D	Boy E	Boy F	Boy G	Boy H	
Boy D	Boy E	Boy F	Boy G	Boy H		
Boy E	Boy F	Boy G	Boy H			
Boy F	Boy G	Boy H				
Boy G	Boy H					
Boy H						

7 + 6 + 5 + 4 + 3 + 2 + 1 = **28**

28

5) Start with the second column. Numbers 6 & 7 are missing-The 7 must be in the 2nd row because there is already a 6, and also the 6 must be in the 3rd row because there is already a 7. Next the diagonal can be filled in with the missing numbers 7 & 8 and the 8 must go in the 2nd row because there is already a 7. That will make Y have to be 5 and X would have to be 6 since the 1st row is missing 6 and 7 and there is already a 7 in that column.
The sum of X + Y or 5 + 6 = **11**

X= 6

Y= 5

X	8	7	5
Y	7	8	6
8	6	5	7
7	5	6	8

X	8		5
Y	7		6
	6		7
	5		8

X	8		5
Y	7	8	6
	6		7
7	5		8

X	8		5
5	7	8	6
	6		7
7	5		8

6	8	7	5
5	7	8	6
	6		7
7	5		8

11

PERENNIAL MATH

TEST # 1

November 2012 Intermediate Level

1) What is the sum of the first five prime numbers?	
2) Waldo has a bag of marbles. There are five red, four blue, three yellow, two white, and one black. How many marbles must he pull out of the bag to be sure that he gets two of the same color?	
3) Mrs. Pickens was sorting the items in the lost and found. She found that there were four times as many gloves as hats and two times as many hats as coats. If there were 88 items in the lost and found, how many gloves were in there?	
4) The numbers given represent the sum of the objects in each row or column. Find the value of one slice of pizza. 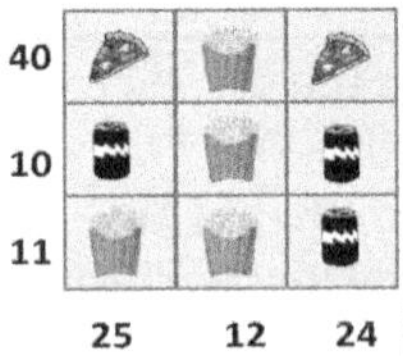	
5) How much is (the sum of the first four multiples of 13 plus the first five multiples of 17) divided by 9?	
6) In this cryptarithm, different letters represent different digits. What number does the letters IN represent? $\begin{array}{r} NO \\ NO \\ NO \\ +\ NO \\ \hline IN \end{array}$	

PERENNIAL MATH

TEST #1

November 2012 Intermediate Level Answers

1) What is the sum of the first five prime numbers?	**28**
2) Waldo has a bag of marbles. There are five red, four blue, three yellow, two white, and one black. How many marbles must he pull out of the bag to be sure that he gets two of the same color?	**6**
3) Mrs. Pickens was sorting the items in the lost and found. She found that there were four times as many gloves as hats and two times as many hats as coats. If there were 88 items in the lost and found, how many gloves were in there?	**64**
4) The numbers given represent the sum of the objects in each row or column. Find the value of one slice of pizza. 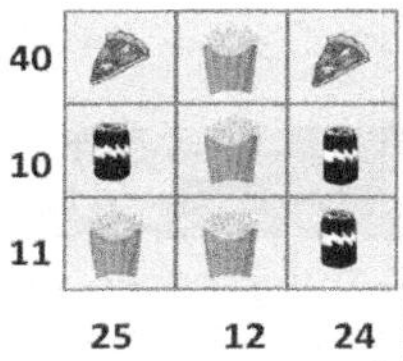	**18**
5) How much is (the sum of the first four multiples of 13 plus the first five multiples of 17) divided by 9?	**50**
6) In this cryptarithm, different letters represent different digits. What number does the letters IN represent? NO + NO + NO + NO = IN	**92**

```
  NO
  NO
+ NO
  NO
----
  IN
```

PERENNIAL MATH

TEST # 1

November 2012 Intermediate Level Solutions

1) The sum of the first five prime numbers is

2 + 3 + 5 + 7 + 11 = **28**

28

2) Since Waldo has a five different colors of marbles in the bag, he could possibly draw out one of each color consecutively. Therefore, on the sixth draw, that sixth marble would have to be the same color of one of the first five marbles drawn, so **it takes 6 draws** for him to be sure that he has two marbles of the same color.

6

3) Make a table to sort the data. There is 2 times as many hats as coats and 4 times as many gloves as hats and the sum of all 3 items is 88. There are **64 gloves**.

Gloves	Hats	Coats	Total
(4 x Hats)	(2 x Coats)		
64	16	8	88

64

4) Use algebra.
The middle column has 3 french fries for 12.
So 12÷3= 4 so french fries are worth 4.
Substituting 4 in for french fries in the first row you have 2 pizzas plus 1 french fry (4) = 40, therefor 40-4 = 36 and the 2 pizzas are 36.
One pizza would = **18**

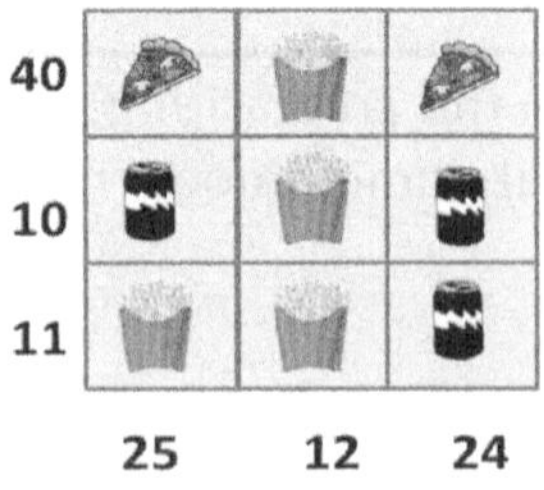

18

PERENNIAL MATH — TEST # 1

November 2012 Intermediate Level Solutions

5) How much is (the sum of the first four multiples of 13 plus the first five multiples of 17) divided by 9?

$$\begin{array}{r} 13 \\ 26 \\ +\quad 39 \\ 52 \\ 65 \\ \hline 195 \end{array} \qquad \begin{array}{r} 17 \\ 34 \\ +\quad 51 \\ 68 \\ 85 \\ \hline 255 \end{array}$$

195+255=450 450 ÷ 9 = **50**.

Observation: Since there are 5 multiples of thirteen there are 1 + 2 + 3 + 4 + 5=15 sets of 13, then you could multiply 13 x 15= 195, and there are 15 sets of 17 or 15 X17= 255. Add 195 + 255= 450 and divide by 9 = 50.

50

6) Since different letters represent different digits in this cryptarithm, then N must be the same throughout. The answer to the addition problem is a 2-digit number so N has to be 1 or 2 so that the number doesn't carry over. Since there are four numbers being added, N cannot be 1 because O x 4 would give an even number when adding up the ones column. Since N must be 2, then O has to be 3 because the only number multiplied by 4 to yield a 2 in the ones-digit is 3 x 4 = 12. So NO = 23.
23 x 4 = 92, therefore, **IN= 92**.

N must be 1 or 2

NO
NO
+ NO
NO
I N

O has to be 3 because 4 x 3 = 12

N has to be **2** because it will be an even number since there are 4 Os that are being added

92

PERENNIAL MATH

TEST #2

December 2012 Intermediate Level

1) Christina sold dance tickets that were consecutively numbered. If the first ticket was 750 and the last ticket she sold was numbered 791, how many tickets did sell?

2) The sum of 2 numbers is 25. Their difference is 7. What is the product of the two numbers?

3) The small rectangles have a perimeter of 16cm and an area of $12cm^2$. How many cm^2 are in the area of the large rectangle?

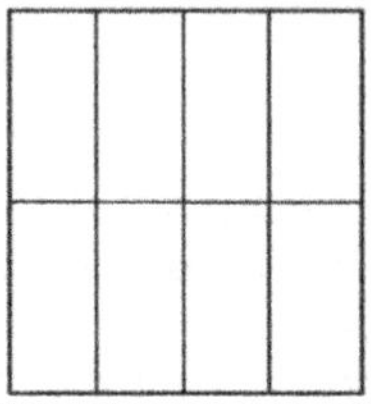

4) Last Halloween there were witches and cats flying over. Clyde noticed there were two and a half times the number of witches as the number of cats and the total of the legs and heads was 75. How many cats were counted?

5) Grady Gizzard was a singer who recorded albums for 15 years. He recorded 3 the first year and each consecutive year recorded 8 more than the number his previous year. Simon Superstar recorded 3 the first year and each consecutive year recorded 1 more than his number the previous year. How many more albums did Grady record compared to Simon?

6) There were 5 snowflakes, 3 candy canes, 2 drummer boys, and 4 stars in a black bag. How many would Cassidy have to pull out to be sure that she pulled out 2 of the same item?

PERENNIAL MATH

TEST #2

December 2012 Intermediate Level

1) Christina sold dance tickets that were consecutively numbered. If the first ticket was 750 and the last ticket she sold was numbered 791, how many tickets did sell?	**42**
2) The sum of 2 numbers is 25. Their difference is 7. What is the product of the two numbers?	**144**
3) The small rectangles have a perimeter of 16cm and an area of $12cm^2$. How many cm^2 are in the area of the large rectangle? 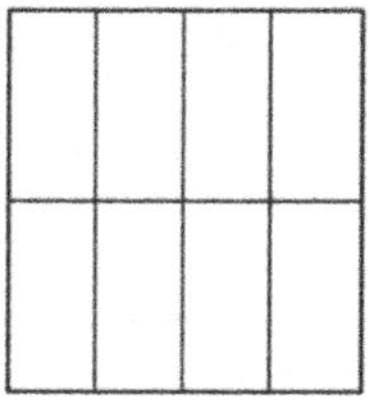	**96**
4) Last Halloween there were witches and cats flying over. Clyde noticed there were two and a half times the number of witches as the number of cats and the total of the legs and heads was 75. How many cats were counted?	**6**
5) Grady Gizzard was a singer who recorded albums for 15 years. He recorded 3 the first year and each consecutive year recorded 8 more than the number his previous year. Simon Superstar recorded 3 the first year and each consecutive year recorded 1 more than his number the previous year. How many more albums did Grady record compared to Simon?	**735**
6) There were 5 snowflakes, 3 candy canes, 2 drummer boys, and 4 stars in a black bag. How many would Cassidy have to pull out to be sure that she pulled out 2 of the same item?	**5**

PERENNIAL MATH

TEST #2

December 2012 Intermediate Level Solutions

1) The first ticket was 750 and the last ticket was 791. You cannot subtract 791-750. You will forget to count ticket 750. If you count ticket 750-759-that is 10 tickets, 760-769 is 10 tickets, 770-779 is 10 tickets, 780-789 is 10 tickets and then there are 2 more 790-791. There are **42** total.

42

2) 2 numbers whose sum is 25 and their difference is 7 are 16 and 9. 16 + 9 = 25, 16-9 = 7. The product of the two numbers 16 x 9 = **144**.

144

3) The small rectangles have a perimeter 16cm and an area of $12cm^2$. The dimensions for a rectangle with 16 cm could only be (1,7)(2,6)(3,5)(4,4). (2,6) are the only dimensions that give an are 6x2 of $12cm^2$. The sides of the large rectangle have lengths of 12 and 8. Therefore, the are of the large rectangle is 12x8=**$96cm^2$**

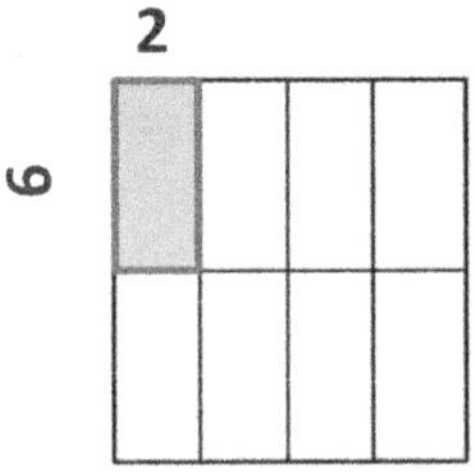

96

4) Make a chart. Witches = 2 ½ x Cats. Legs + Heads = 75Last Halloween there were witches and cats flying over. Heads =Witches + Cats. Legs = (Witches x 2) + (Cats x 4). There were **6 Cats**.

Witches	**15**
Cats	**6**
Heads	**21**
Legs	**54**
Total Heads + Legs (75)	**75**

6

PERENNIAL MATH TEST #2

December 2012 Intermediate Level Solutions

5) Make a table to compare the recordings and add up the sum for both Grady and Simon. Subtract Simon's total from Grady's total. Grady recorded 3 the first year, then 8 more than the previous year for each consecutive year. Simon recorded 3 the first year and 1 more than the previous year each consecutive year. Grady recorded a total of 885. Simon recorded a total of 150. 885-150= **735**

																Total
Grady	3	11	19	27	35	43	51	59	67	75	83	91	99	107	115	885
Simon	3	4	5	6	7	8	9	10	11	12	13	14	15	16	17	150

735

6) Act it out.

There were 5 snowflakes, 3 candy canes, 2 drummer boys, and 4 stars in a black bag. She could pull out a snowflake, a candy cane, a drummer boy, and a star. Then the next item would be one of the four items and she would for sure have 2 of some item. Cassidy would have to pull out **5 items** to be sure that she pulled out 2 of the same item.

5

PERENNIAL MATH

TEST #3

January 2013 Intermediate Level

1) What is the average of the first ten odd whole numbers?

2) If you have two types of bread, four types of cheese and three kinds of meat, how many different types of sandwiches can you make if you must use only one type of bread, cheese , and meat on your sandwich?

3) What is the largest two-digit number that is twice the product of its digits?

4) This 4x4x4 cube had all six faces of the cube painted blue. Suppose the cube is then cut into 1x1x1 cubes along the lines shown. How many 1x1x1 cubes will have blue paint on just two faces?

5) If Joey puts his baseball cards in stacks of 5 there are no cards left over. If he puts his cards in stacks of 4, there are 2 cards left over. If he puts his cards in stacks of 3, there is 1 card left over. What is the fewest number of cards that Joey can have if he has more than 50 cards?

6) Place the numbers 1,2,3,4, and 6 in the circles so that the sum of the bottom three numbers is 1 more that the sum of the right 3 numbers and two more than the sum of the left 3 numbers. What is the sum of the bottom 3 numbers?

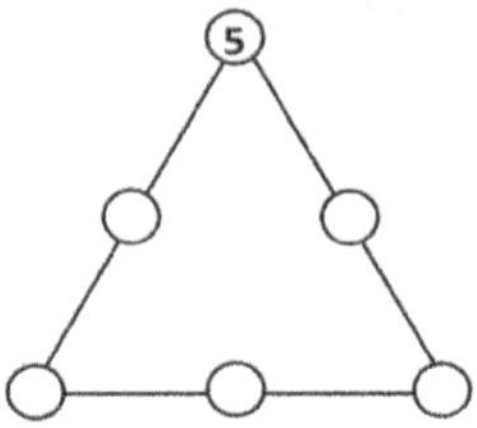

PERENNIAL MATH

TEST #3

January 2013 Intermediate Level Answers

1) What is the average of the first ten odd whole numbers?	**10**
2) If you have two types of bread, four types of cheese and three kinds of meat, how many different types of sandwiches can you make if you must use only one type of bread, cheese , and meat on your sandwich?	**24**
3) What is the largest two-digit number that is twice the product of its digits?	**36**
4) This 4x4x4 cube had all six faces of the cube painted blue. Suppose the cube is then cut into 1x1x1 cubes along the lines shown. How many 1x1x1 cubes will have blue paint on just two faces? 	**24**
5) If Joey puts his baseball cards in stacks of 5 there are no cards left over. If he puts his cards in stacks of 4, there are 2 cards left over. If he puts his cards in stacks of 3, there is 1 card left over. What is the fewest number of cards that Joey can have if he has more than 50 cards?	**70**
6) Place the numbers 1,2,3,4, and 6 in the circles so that the sum of the bottom three numbers is 1 more that the sum of the right 3 numbers and two more than the sum of the left 3 numbers. What is the sum of the bottom 3 numbers? 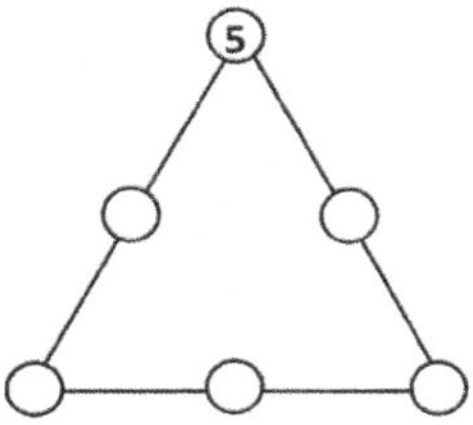	**13**

PERENNIAL MATH

TEST # 3

January 2013 Intermediate Level Solutions

1) You could average the first ten odd whole numbers:
1 + 3 + 5 + 7 + 9 + 11 + 13 + 15 + 17 +19 = 100÷10 = 10
Or you could notice that the beginning and ending numbers are pairs that add up to 20 each (1 + 19) ÷ 2 = 10, (3 + 17)) ÷ 2 = 10, etc..

10

2) Using the principles of multiplication - If you have two types of bread, four types of cheese and three kinds of meat, you can make 24 different types of sandwiches if you must use only one type of bread, cheese , and meat on your sandwich. [2 x 4 x 3] Or you could draw it out to establish a pattern

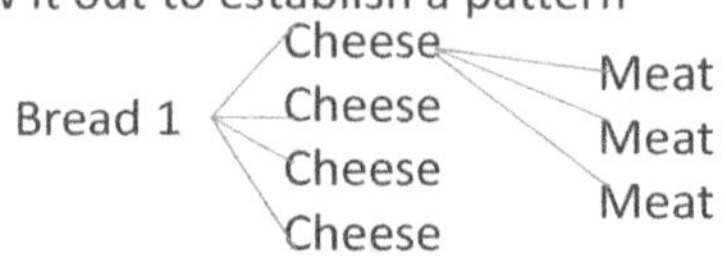

Bread1 Cheese1Meat1
Bread1 Cheese1Meat2
Bread1 Cheese1Meat3

This occurs for each cheese, so since there are 4 cheeses, there are 12 sandwiches for each bread, and since there are 2 breads there is a total of 2 x 12 = 24 sandwiches.

24

3) Since the number we are looking for is twice the product of its digits, it has to be an even number. Also, none of the two digit numbers ending in zero are a possibility because their product would be zero. 3 x 6 x 2 = 36, So 36 is the answer. No other even two-digit numbers work out to be twice the product of their digits. The chart below demonstrates the products.

Number	Tens Digit	Ones Digit	x2	Product
12	1	2	2	4
14	1	4	2	8
16	1	6	2	12
18	1	8	2	16
22	2	2	2	8
24	2	4	2	16
26	2	6	2	24
28	2	8	2	32
32	3	2	2	12
34	3	4	2	24
36	3	6	2	36
38	3	8	2	48
42	4	2	2	16
44	4	4	2	32
46	4	6	2	48
48	4	8	2	64

Number	Tens Digit	Ones Digit	X2	Product
52	5	2	2	20
54	5	4	2	40
56	5	6	2	60
58	5	8	2	80
66	6	6	2	72
68	6	8	2	96
72	7	2	2	28
74	7	4	2	56
76	7	6	2	84
78	7	8	2	112
88	8	8	2	128
92	9	2	2	36
94	9	4	2	72
96	9	6	2	108
98	9	8	2	144

36

PERENNIAL MATH

TEST # 3

January 2013 Intermediate Level Solutions

4) If this 4x4x4 cube had all six faces of the cube painted blue and was cut into 1x1x1 cubes along the lines, then there would be 24 cubes with blue paint on just two faces. Each side of the cube would have 8 of its 1 x 1 x 1 cubes with blue paint only on 2 sides. If the cube is sliced on the lines, the front and back layers both have 8 cubes with blue paint on 2 sides and the middle 2 layers both have 4 cubes each with paint only on 2 sides, 8 + 8 + 4 + 4= **24 cubes with paint only on 2 sides**.

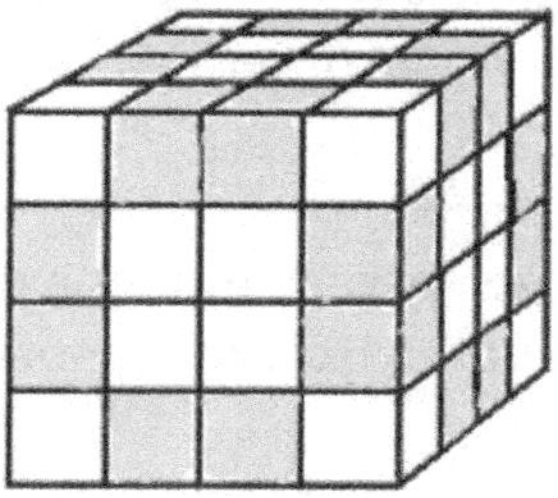

24

5) Since there are no cards left over when Joey puts his baseball cards in stacks of 5, then the number is a multiple of 5 which is greater than 50. Start with 55, it cannot be the answer because it has a remainder of 3 when divided by 4. 60 is not the answer because there is no remainder when divided by 4. 65 is not the answer because the is only a remainder of 1 when dividing by 4. **70** is the answer because when he puts his cards in stacks of 4, there is 2 left over and when he puts his cards in stacks of 3, there is 1 card left over.

$$70 \div 4 = 18 \text{ with remainder of } 2$$
$$70 \div 3 = 23 \text{ with remainder of } 1$$

70

6) Place the number 1, 2, 3, 4, and 6 so that the sum of the bottom three numbers is 1 more that the sum of the right 3 numbers and two more than the sum of the left 3 numbers. **The sum of the bottom 3 numbers is 13**. The sum of the right three numbers is 12 and the sum of the left three numbers is 11.

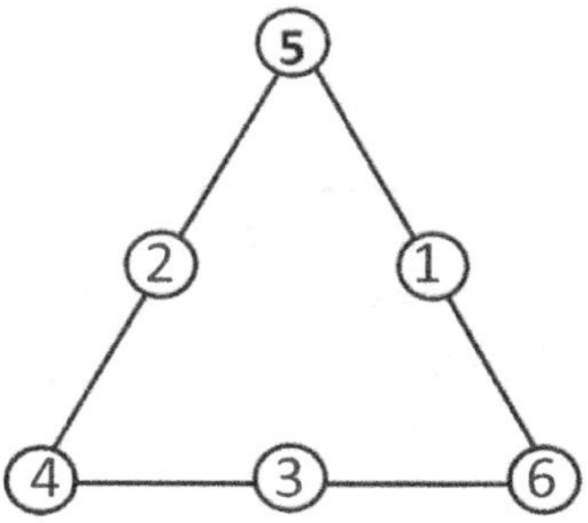

13

PERENNIAL MATH

TEST #4

February 2013 Intermediate Level

1) What is the smallest whole number you can make whose digits are all different numbers and the sum of the digits is 21?

2) Seven people met at a party and shook hands once with each other. How many total handshakes were exchanged?

3) Arrange the numbers 1 to 9 in the squares so that the sum of each row, column, and diagonal is 15. What is the middle number?

4) Mrs. Waldrup's class conducted a survey and discovered that 2 out of 5 students do not own a cell phone. There were 27 students who had a cell phone. How many students participated in the survey?

5) Cathy, Suzie, and Mary went shopping. They counted the money in their purses and found that Cathy and Suzie together had $57, Suzie and Mary had $49, and Cathy and Mary had $64. How many dollars did Mary have?

6) In a class of 200 students, 112 students are in the band, 94 students are on sports teams, and 30 students do both activities. How many students do not do band or sports?

PERENNIAL MATH

TEST #4

February 2013 Intermediate Level Answers

1) What is the smallest whole number you can make whose digits are all different numbers and the sum of the digits is 21?	**489**
2) Seven people met at a party and shook hands once with each other. How many total handshakes were exchanged?	**21**
3) Arrange the numbers 1 to 9 in the squares so that the sum of each row, column, and diagonal is 15. What is the middle number? 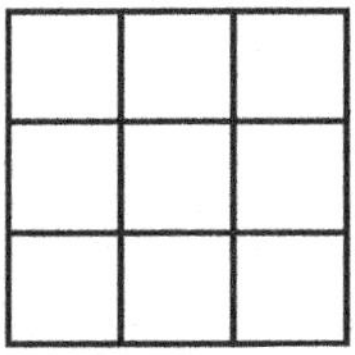	**5**
4) Mrs. Waldrup's class conducted a survey and discovered that 2 out of 5 students do not own a cell phone. There were 27 students who had a cell phone. How many students participated in the survey?	**45**
5) Cathy, Suzie, and Mary went shopping. They counted the money in their purses and found that Cathy and Suzie together had \$57; Suzie and Mary had \$49; and Cathy and Mary had \$64. How many dollars did Mary have?	**28**
6) In a class of 200 students, 112 students are in the band, 94 students are on sports teams, and 30 students do both activities. How many students do not do band or sports?	**24**

PERENNIAL MATH

TEST #4

February 2013 Intermediate Level Solutions

1) The least whole number whose digits are all different numbers that add up to 21 is 489. If 9 is not used as one of the numbers, the next smallest number would be 678. So when 9 is used, the next smallest number that can be used that will not allow a duplicate and will allow a sum of 21 is the number 4. (The numbers 1, 2, and 3 will not work because they don't add up to 21 when combined with numbers or they force you to use 9 again). The numbers 4 + 8 + 9 add up to 21 and 4 would be the lowest number to be placed in the hundreds column.

489

2) Seven people met at a party and shook hands once with each other. There were a total of 6 +5 + 4 + 3 + 2 + 1= 21
handshakes exchanged.

Person 1	Person 2	Person 3	Person 4	Person 5	Person 6	Person 7
Person 2	Person 3	Person 4	Person 5	Person 6	Person 7	0
Person 3	Person 4	Person 5	Person 6	Person 7	1	
Person 4	Person 5	Person 6	Person 7	2		
Person 5	Person 6	Person 7	3			
Person 6	Person 7	4				
Person 7	5					
6						

21

3) If you must take 3 (different) digits out of the digits 1 up to 9, which total 15 each time, then there are the following possibilities:

1+5+9
1+6+8
2+4+9
2+5+8
2+6+7
3+4+8
3+5+7
4+5+6.

If we count the appearance of the 1, 2, 3, 4, 5, 6, 7, 8 and 9 in the 8 possibilities, then we get the following result:

1 = 2x
2 = 3x
3 = 2x
4 = 3x
5 = 4x
6 = 3x
7 = 2x
8 = 3x
9 = 2x

This helps with the placement of the numbers since 5 appears 4x, it has to go in the middle, the numbers that appear 3x must go on the corners. This is one possible arrangement below:

2	7	6
9	5	1
4	3	8

5

PERENNIAL MATH

TEST #4

February 2013 Intermediate Level Solutions

4) 2 out of 5 students do not own a cell phone, so 3 out of every 5 must own a cell phone. There were 27 students who had a cell phone. How many students participated in the survey?

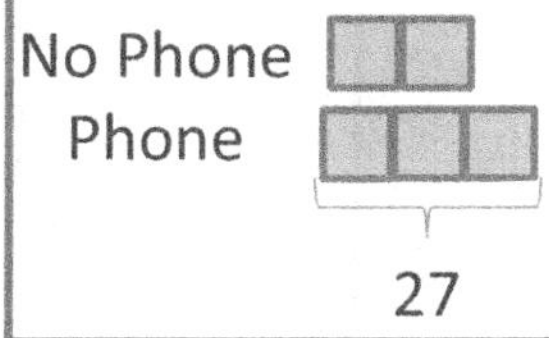

3 portions = 27, so 1 portion is = 9 students.
Therefore, **45 participated** in the survey

45

5) Use Algebra:

$$\begin{aligned} C + S \quad\quad\quad &= 57 \\ S + M &= 49 \\ C + \quad\quad M &= 64 \end{aligned}$$

2C + 2S + 2M = 170 (Now divide both sides by 2)
C + S + M = 85. Now you can substitute C + S in the equation and subtract.
(C + S + M) – (C + S) = 85 – 57 = 28, so M or **Mary had $28**.

28

6) In a class of 200 students, 112 students are in the band, 94 students are on sports teams, and 30 students do both activities. Region I is band only, Region II is for sports only. Region III is both activities. Region IV is neither activity. Band only is (112 – 30) = 82 students. Sports only is (94 – 30) = 64. Region I + II + III + IV = 200. 82 +64 +30 + IV= 200. **Region IV = 24**.

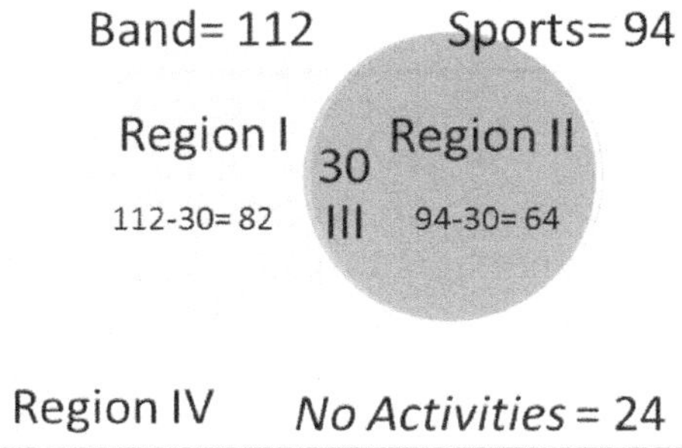

24

PERENNIAL MATH

TEST #1

November 2013 Intermediate Level

1) Two cups of flour are needed to bake half a dozen cupcakes. How many cups of flour are needed for 45 cupcakes?	
2) If you roll 2 dice with numbers 1 through 6 on the sides, how many different sum totals could you get adding the numbers together on each face?	
3) Simplify: $\frac{(11 \times 15)+(2 \times 5)^2+10}{3^2+4^2}$	
4) Molly bought a sucker for 35 cents. How many different ways could she have paid for it with coins from her piggy bank if she had dimes, nickels, and pennies?	
5) Find the area of the figure below: 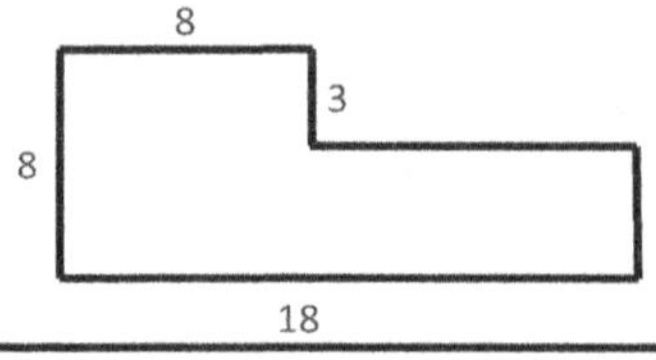	
6) Sharon wants to get an A in English. On her last test she had an 87 and her previous scores were 92, 98, 83, 81, 94. What score will she need to make on her next test to average a 90 in English?	

PERENNIAL MATH

TEST # 1

November 2013 Intermediate Level Answers

1) Two Cups of flour are needed for half a dozen cupcakes. How many cups of flour are needed for 45 cupcakes?	**15**
2) If you roll 2 dice with numbers 1 through 6 on the sides, how many different sum totals could you get adding the numbers together on each face?	**11**
3) Simplify: $\frac{(11 \times 15)+(2 \times 5)^2+10}{3^2+4^2}$	**11**
4) Molly bought a sucker for 35 cents. How many different ways could she have paid for it with coins from her piggy bank if she had dimes, nickels, and pennies?	**20**
5) Find the area of the figure below: 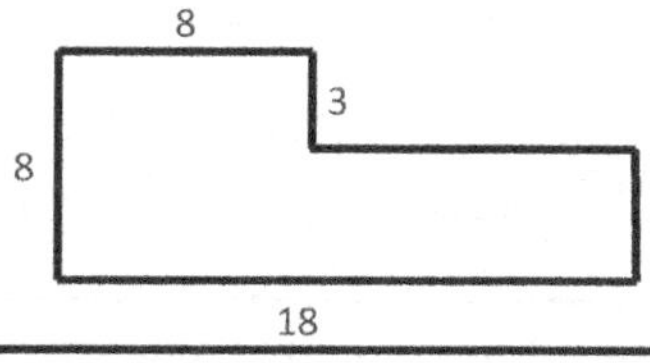	**114**
6) Sharon wants to get an A in English. On her last test she had an 87 and her previous scores were 92, 98, 83, 81, 94. What score will she need to make on her next test to average a 90 in English?	**95**

PERENNIAL MATH

TEST # 1

November 2013 Intermediate Level Solutions

1) 2 cups of flour = 6 cupcakes. You have to make a total of 45 cupcakes. Divide 45 by 6 and then multiply it by 2.

Equation:
2 cups = 6 cupcakes

45 cupcakes ÷ 6 = 7.5
7.5 x 2 = **15 cups of flour**

15

2) 2 Dice, you need to find how many different sums there will be, not the same sums. So you know that the lowest number that it will add up to is 2, and the highest number is 12, so you only count each sum once, so you would count from the sum of 1+1= 2 and #1 and then count up to the last/highest sum. The total different sums would be **11.**

+	1	2	3	4	5	6
1	2	3	4	5	6	7
2	3	4	5	6	7	8
3	4	5	6	7	8	9
4	5	6	7	8	9	10
5	6	7	8	9	10	11
6	7	8	9	10	11	12

11

3) Simplify: $\frac{(11 \times 15)+(2\times 5)^2+10}{3^2+4^2} = \frac{(165)+(10)^2+10}{9+16} =$

$$\frac{165+100+10}{25} = \frac{275}{25} = \mathbf{11}$$

11

4) You will need to list out all of the different ways you can make 35 cents from Dimes, Nickels and/or Pennies. Let's start with Dimes:

1) 3 Dimes, 1 Nickel
2) 3 Dimes, 5 Pennies
3) 2 Dimes, 3 Nickels
4) 2 Dimes, 2 Nickels, 5 Pennies
5) 2 Dimes, 1 Nickel, 10 Pennies
6) 2 Dimes, 15 Pennies
7) 1 Dime, 5 Nickels
8) 1 Dime, 4 Nickels, 5 Pennies
9) 1 Dime, 3 Nickels, 10 Pennies
10) 1 Dime, 2 Nickels, 15 Pennies
11) 1 Dime, 1 Nickel, 20 Pennies
12) 1 Dime, 25 Pennies
13) 7 Nickels
14) 6 Nickels, 5 Pennies
15) 5 Nickels, 10 Pennies
16) 4 Nickels, 15 Pennies
17) 3 Nickels, 20 Pennies
18) 2 Nickels, 25 Pennies
19) 1 Nickel, 30 Pennies
20) 35 Pennies

20

PERENNIAL MATH — TEST #1

November 2013 Intermediate Level Solutions

5) You will need to find the Area for two shapes, a square and a rectangle. Put a invisible line to complete the square, this will separate and give you the two shapes. A square has 4 equal sides, so you have to find the length and width of the rectangle. So you know that each side of the square are 8 inches long, so now, you need to find the remaining inches on the side where the square and rectangle connect: x + 3 = 8, and then find the remaining inches on the bottom length: x + 8 = 18. solve for x on both. Once you have both sides, compute the area for both the square and rectangle and then add the results together to get total area.

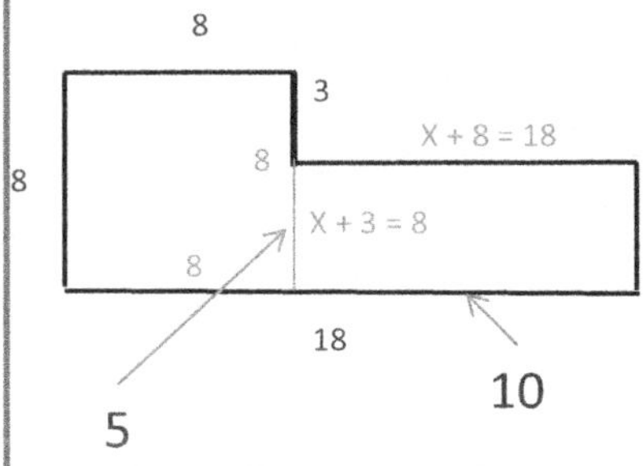

Step 1 **Rectangle Equation**	*Step 2* **Area Equation for Both**	*Step 3* **Total Area**
Short Side: **X + 3 = 8**	A = L x W	Add both Areas Together.
X = 8 – 3 **X =** 5	Square: 8 x 8 = **64**	**64** + 50 = 114
		Total Area = 114
Long Side: **X + 8 = 18** **X = 18- 8** **X =** 10	Rectangle: 5 x 10 = 50	

114

6) In order to solve the problem you will need to find Sharon's current average and subtract it from the average she wants to obtain. First you will have to look at her average in terms of the sum total points earned. Using the formula Average=sum of grades/number of grades, Sharon's current average is 87 + 92 + 98 + 83 + 81 + 94=535/6 or 89.17 In order for her to get an average of 90 she would now have 7 grades and 7grades multiplied by 90 is 630 total points that she must earn. So 630 points minus the current 535 points is 95. Sharon will **have to make a 95 on her last test** to get an A average in English.

95

PERENNIAL MATH

TEST #2

December 2013 Intermediate Level

1) If you multiply a number N by 28 and then subtract 8, you end up with 160. What is the number N?

2) Stanley signed 1/2 of the holiday cards, and Regina signed 1/3 of them. If there are 48 cards in all, how many are left to be signed?

3) Place the numbers 1-9 in the squares so that each set of 4 numbers surrounding each ornament adds up to 17. What number will N be?

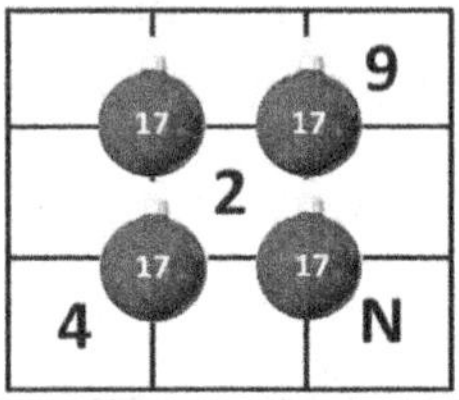

4) On the way to the beach Bryan drove for 1/2 of the trip, and Darrell drove for 1/4 of the trip. Jasmine and Holly divided the rest of the driving evenly between them. If the entire trip was 168 miles, how many miles did Holly drive?

5) The number of boys at the pool was six more than twice the number of girls. Altogether there were 156 swimmers. How many were boys?

6) One half of the people wore team colors at the game. Two thirds of those people wore team pins as well. One fourth of those with team colors and team pins had pom poms. Twenty-five people had team colors and pom poms, but no pins. One hundred people had only pom poms. If there were 1,824 people at the game, how many had pom poms?

PERENNIAL MATH TEST #2

December 2013 Intermediate Level Answers

1) If you multiply a number N by 28 and then subtract 8, you end up with 160. What is the number N?	**6**
2) Stanley signed 1/2 of the holiday cards, and Regina signed 1/3 of them. If there are 48 cards in all, how many are left to be signed?	**8**
3) Place the numbers 1-9 in the squares so that each set of 4 numbers surrounding each ornament adds up to 17. What number will N be? 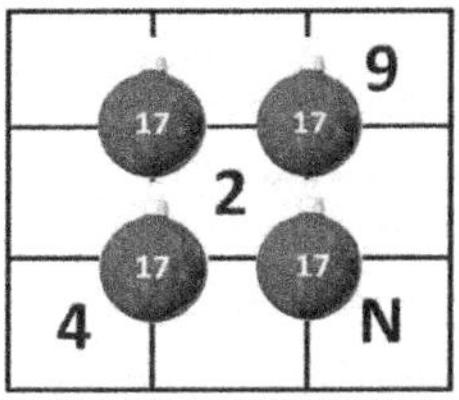	**7**
4) On the way to the beach Bryan drove for 1/2 of the trip, and Darrell drove for 1/4 of the trip. Jasmine and Holly divided the rest of the driving evenly between them. If the entire trip was 168 miles, how many miles did Holly drive?	**21**
5) The number of boys at the pool was six more than twice the number of girls. Altogether there were 156 swimmers. How many were boys?	**106**
6) One half of the people wore team colors at the game. Two thirds of those people wore team pins as well. One fourth of those with team colors and team pins had pom poms. Twenty-five people had team colors and pom poms, but no pins. One hundred people had only pom poms. If there were 1,824 people at the game, how many had pom poms?	**277**

PERENNIAL MATH **TEST #2**

December 2013 Intermediate Level Solutions

1) If you multiply a number N by 28 and then subtract 8, you end up with 160.

$28N - 8 = 160$

$28N = 168$

$N = 168 \div 28$

$\mathbf{N = 6}$

6

---

2) Stanley signed ½ or (24) of the 48 holiday cards, and Regina signed 1/3 or (16) of them. 24 + 16 = 40 signed cards. 48 – 40 = **8 cards left to be signed**. **8**

---

3) Since 9 and 2 are given in the top right squares, the remaining 2 squares have to add up to 6 to make the sum around the ornament = 17. The only possible numbers to use are 1 and 5 because 2 and 4 are already used. Now the bottom right squares are 4 and 2 so the other two squares around them add up to 11. The only numbers left to use are 8 and 3 because 9, 2, and 5, 6 combinations are not available. Since we know that 8 and 3, 1 and 5, are to be used then the remaining corners must be 6 and 7. This helps us eliminate where to place the numbers and once we play around with them, then the solution will appear. The bottom right corner is 7 so **N = 7**.

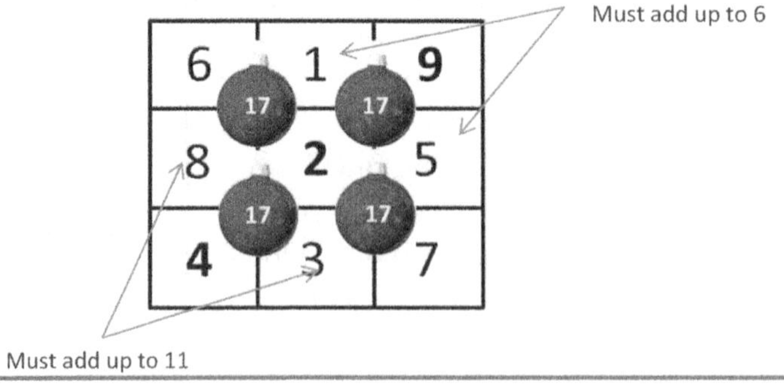

7

PERENNIAL MATH TEST #2

December 2013 Intermediate Level Solutions

4) Bryan drove 1/2 x 168 = 84 miles
Darrell drove 1/4 x 168= 42 miles
84 + 42 = 126 miles driven
168 - 126 = 42 miles left
42 ÷ 2 = **21 miles driven by Holly**

21

---

5) Use Algebra. The number of boys at the pool was six more than twice the number of girls. Altogether there were 156 swimmers. Let x represent the number of girls.

x + (2x + 6) =156
3x = 150
x = 50
So there are 2x + 6 = **106 boys**.

106

---

6) 1824 ÷ 2 = 912 with team colors

2/3 x 912 = 608 with pins and team colors

1/4 x 608 = 152 with pins, team colors, and pom poms

152 + 100 + 25 = **277 with pom poms**

277

PERENNIAL MATH

TEST #3

January 2014 Intermediate Level

1) 17 x 36 = 42 + ☐

2) What digit does the star represent in this problem?

$$\begin{array}{r} ★★ \\ \times \quad ★ \\ \hline 704 \end{array}$$

3) Jonah was driving and started at mile marker 344. He got off at mile marker 392. What would be the mile marker number of one fourth of the way?

4) Miriam played a game where she could earn 3 points or 16 points each turn. What is the maximum amount of turns that she could score 16 points when she totaled 148 points?

5) Sally went to a store and spent one-forth of her money. She went to a second store where she spent one-third of what remained, and then had $12 when she left. How many dollars did she have to begin with at the first store?

6) Amy and Kayla each threw darts at the target. Amy hit region X four times and region Y two times for a score 54 points. Kayla hit region X three times and region Y three times for a score of 60 points. How many points was region X worth?

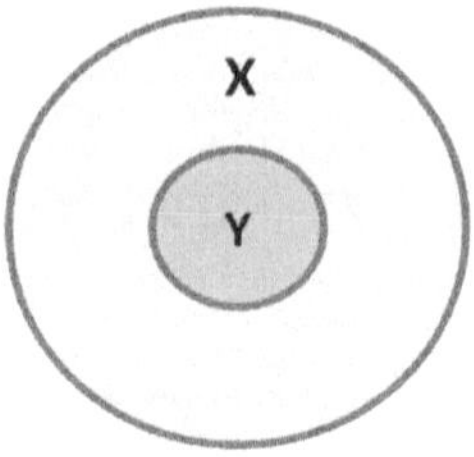

PERENNIAL MATH

TEST #3

January 2014 Intermediate Level Answers

1) 17 x 36 = 42 + ☐	**570**
2) What digit does the star represent in this problem? ★★ x ★ 704	**8**
3) Jonah was driving and started at mile marker 344. He got off at mile marker 392. What would be the mile marker number of one fourth of the way?	**356**
4) Miriam played a game where she could earn 3 points or 16 points each turn. What is the maximum amount of turns that she could score 16 points when she totaled 148 points?	**7**
5) Sally went to a store and spent one-forth of her money. She went to a second store where she spent one-third of what remained, and then had $12 when she left. How many dollars did she have to begin with at the first store?	**24**
6) Amy and Kayla each threw darts at the target. Amy hit region X four times and region Y two times for a score 54 points. Kayla hit region X three times and region Y three times for a score of 60 points. How many points was region X worth? 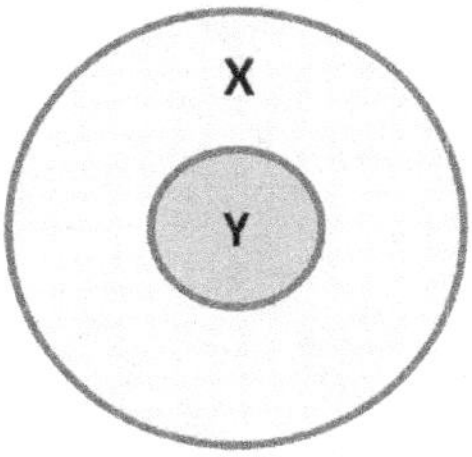	**7**

PERENNIAL MATH

TEST #3

January 2014 Intermediate Level Solutions

1) 17 x 36 = 42 + □

17 x 36 = 612

612 – 42 = **570**

570

2) Since the star represents the same digit each time, then star x star yields a 4. The only digits that could do that are 2 and 8. Two does not work because the final product would be 44, but 8 works because it carries and the product of 88 x 8 is **704**.

★★
x ★
704

8

3) First find the distance of the trip: 396 – 344 = 48 miles. One fourth of the way would be 48/4 = 12 miles. The mile marker number would be 344 + 12 = **356**.

356

PERENNIAL MATH

TEST # 3

January 2014 Intermediate Level Solutions

4) Make a chart to see how many of each point combinations will give a total of 148 points and start with the most amount of 16 points possible. Note: since the total point value is even, there must be an even number of 3 points scored. There were twelve 3 points scored and **seven turns** where 16 points were scored.

3 points	16 points	Total Points
2	9	150
4	8	140
6	8	146
8	8	152
10	7	142
12	7	148

7

5) Sally spent 1/4 of her money in store 1 and had 3/4 left. In the second store she spent 1/3 of 3/4 of her original amount. In both stores she spent 1/4+(1/3 x 3/4) 1/4 + 3/12 = 1/4 + 1/4 or ½ of her money. Clearly she had 1/2 of his money left which was equal to $12.She originally had $24.

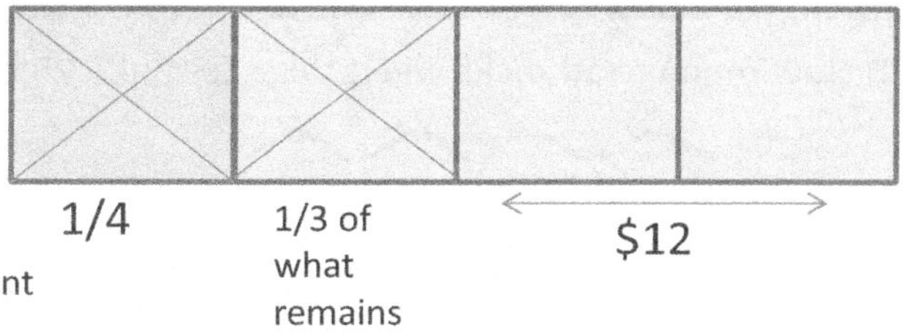

Algebraic Solution:
Let x represent the original amount
$x*(3/4)*(2/3)=12$
Multiply fractions
$(6/12)*x=12$
$(1/2)*x=12$
Multiply both sides by 2
x=24

$24

6) Use algebra: (Kayla) $4x + 2y = 54$ Amy $3x + 3y = 60$
$3x + 3y = 60$ divide both sides by 3 and you get $x + y = 20$; therefore $2x + 2y = 40$.

Subtract this from Kayla's $4x + 2y = 54$

$$\begin{array}{r} 4x + 2y = 54 \\ \underline{-2x + 2y = 40} \\ 2x \quad\quad = 14 \\ \mathbf{x = 7} \end{array}$$

7

PERENNIAL MATH TEST #4

February 2014 Intermediate Level

1) A box is made up of pine and oak. The ratio of pine to oak is 3: 8, respectively. If the box weighs 33 pounds, how many pounds does the oak weigh?

2) How many numbers between 10 and 300 have the sum of their digits equal to 9?

3) Pham read a chapter that had 11 pages. The sum of the page numbers is 187. What is the first page of the chapter?

4) How many toothpicks will it take to make 25 houses?

5) John has one third as many cards as Barry. Barry has one third as many as Steve. Steve has 9 times as many as Tim. Tim has 8 cards. How many cards do John and Tim have altogether?

6) Maurice lists four consecutive multiples of some number. The average of the first two multiples is 15 and the average of the last two is 27. What is the sum of all four multiples?

PERENNIAL MATH

TEST #4

February 2014 Intermediate Level Answers

1) A box is made up of pine and oak. The ratio of pine to oak is 3: 8, respectively. If the box weighs 33 pounds, how many pounds does the oak weigh?	**24**
2) How many numbers between 10 and 300 have the sum of their digits equal to 9?	**26**
3) Pham read a chapter that had 11 pages. The sum of the page numbers is 187. What is the first page of the chapter?	**12**
4) How many toothpicks will it take to make 25 houses? 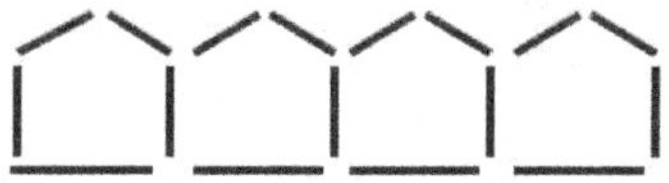	**101**
5) John has one third as many cards as Barry. Barry has one third as many as Steve. Steve has 9 times as many as Tim. Tim has 8 cards. How many cards do John and Tim have altogether?	**16**
6) Maurice lists four consecutive multiples of some number. The average of the first two multiples is 15 and the average of the last two is 27. What is the sum of all four multiples?	**84**

PERENNIAL MATH

TEST # 4

February 2014 Intermediate Level Solutions

1) Since the ratio is 3:8, there are 11 total parts , thus you need to set up a ratio comparing oak to total wood. Since this ratio will be proportional to the make up of the 33 pound box, you set up and solve as follows...

$$\frac{8\ pounds\ oak}{11\ pounds\ total} = \frac{x\ pounds\ oak}{33\ pounds\ total}$$

CROSS MULTIPLY

11x = 264

X= 264/11

X=24 or 24 pounds of oak

24

2) Make an ordered list:

18	108	207
27	117	216
36	126	225
45	135	234
54	144	243
63	153	252
72	162	261
81	171	270
90	180	

There are 26 total numbers

26

3) n= first page read page number

Since she read 11 pages and the sum of the page numbers was 187, set up an equation: n+(n+1)+(n+2)+......+(n+10)=187

Since there were 11 pages, there are 11n

11n+1+2+....10 = 187

11n+55=187

11n=187-55

11n=132

n=132/11

n=12 the **first page number was 12**

12

PERENNIAL MATH

TEST # 4

February 2014 Intermediate Level Solutions

4) The pattern shows that the first toothpick house is made of five toothpicks. Each time a house is added, four more toothpicks are added.
5+4n
n=toothpick houses added
n=24, since we are looking for the amount of toothpicks after 25 houses (1 house is already present)
5+ 4(24) = 101 (**101 toothpicks present**)

101

---

5) Tim has 8 cards
Steve has 8 * 9 cards = 72 cards (Steve has 9 times as many as Tim.)
Barry has 72 ÷ 3 cards = 24 cards (Barry has one third as many as Steve.)
John has 24 ÷ 3 cards = 8 cards (John has one third as many cards as Barry.)

John + Tim = 8 cards + 8 cards = **16 cards**

16

---

6) Since the average of the last two multiples is 27, then it is the midpoint and the two multiples must add to 54.
The average of the first two multiples is 15, then it is the midpoint and the two multiples must add up to 30.
The four multiples are: 12, 18, 24, 30
The **sum of these multiples is 84**

84

Made in the USA
Monee, IL
13 February 2026

44134413R00036